GAINST NATURE

UNTIMELY MEDITATIONS

THE MIT PRESS
CAMBRIDGE, MASSACHUSETTS
LONDON, ENGLAND

AGAINST NATURE

LORRAINE DASTON

© 2019 Massachusetts Institute of Technology

First published in the series "De Natura" (edited by Frank Fehrenbach), which is part of *Fröhliche Wissenschaft* at Matthes & Seitz Berlin: © MSB Matthes & Seitz Berlin Verlagsgesellschaft mbH, Berlin 2018. All rights reserved.

This book was set in PF DinText Pro by Toppan Best-set Premedia Limited. Printed and bound in the United States of America.

Library of Congress Cataloging-in-Publication Data

Names: Daston, Lorraine, 1951– author.
Title: Against nature / Lorraine Daston.
Other titles: Gegen die Natur. English
Description: Cambridge, MA : MIT Press, 2019. | Series: Untimely meditations ; 17 | Includes bibliographical references.
Identifiers: LCCN 2018049313 | ISBN 9780262537339 (pbk. : alk. paper)
Subjects: LCSH: Philosophical anthropology. | Philosophy of nature. | Ethics.
Classification: LCC BD450 .D32513 2019 | DDC 113—dc23 LC record available at https://lccn.loc.gov/2018049313

10 9 8 7 6 5 4 3 2 1

CONTENTS

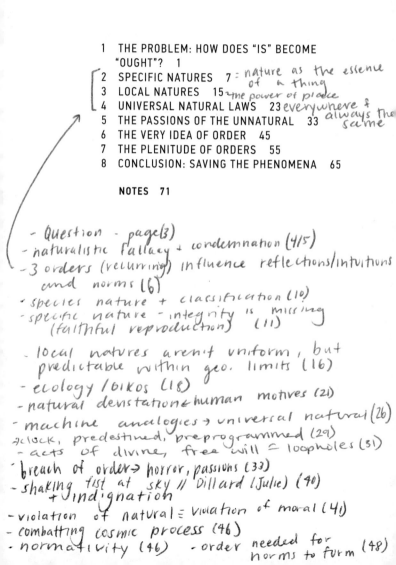

= nature as the essence of a thing

the power of place

everywhere & always the same

- Question - page(3)
- naturalistic fallacy + condemnation (4/5)
- 3 orders (recurring) influence reflections/intuitions and norms (6)
- species nature + classification (10)
- specific nature - integrity is missing (faithful reproduction) (11)
- local natures aren't uniform, but predictable within geo. limits (16)
- ecology / oikos (18)
- natural deviation ← human motives (21)
- machine analogies → universal natural (26)
- clock, predestined, preprogrammed (29)
- acts of divine, free will = loopholes (31)
- breach of order → horror, passions (33)
- shaking fist at sky // Dillard (Julie) (40) + indignation
- violation of natural = violation of moral (41)
- combatting cosmic process (46)
- normativity (46) - order needed for norms to form (48)

human impulse to make nature meaningful (60)

1 THE PROBLEM: HOW DOES "IS" BECOME "OUGHT"?

In his *Anthropology from a Pragmatic Point of View* (1798), Immanuel Kant remarked: "It is noteworthy that we can think of no other suitable form for a rational being than that of a human being. Every other form would represent, at most, a symbol of a certain quality of the human being—as the serpent, for example, is an image of evil cunning—but not the rational being himself. Therefore we populate all other planets in our imagination with nothing but human forms, although it is probable that they may be formed very differently, given the diversity of the soil that supports and nourishes them, and the different elements of which they are composed."[1] The many depictions of the serpent with a human head who corrupted Adam and Eve implicitly make Kant's point: a serpent who could speak and reason so beguilingly was as much person as reptile (fig. 1). Although Kant was firmly convinced of the existence and physical diversity of nonhuman rational beings, he assumed that this diversity made no difference to their character as *rational* beings: whether they were rational Martians or rational angels, reason was reason everywhere in the universe.[2] I would like to offer an alternative to this brand of Kantian philosophical anthropology: it matters to reason—not just to sensibility and psychology—what kind of species we are. The kind of philosophical anthropology I am proposing is an inquiry into *human* reason, rather than universal Reason *tout court*.

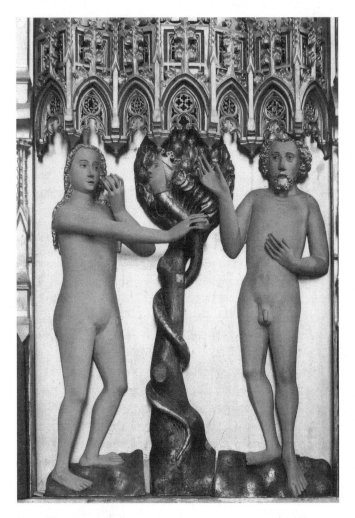

Figure 1
Anonymous Master, *Adam and Eve in Paradise* (ca. 1370),
Doberan Cathedral, Bad Doberan, Germany.

This project makes sense only when anchored in a genuine problem, one of sufficient historical and cultural generality to be a plausible candidate for a philosophical anthropology (as opposed to a cultural anthropology or a history of a particular time and place). The question I would like to address can be simply posed: Why do human beings, in many different cultures and epochs, pervasively and persistently, look to nature as a source of norms for human conduct? Why should nature be made to serve as a gigantic echo chamber for the moral orders that humans make? It seems superfluous to duplicate one order with another, and highly dubious to derive the legitimacy of the human order from its alleged original in nature. Yet in ancient India and in ancient Greece, in medieval France and Enlightenment America, in the latest controversy over homosexual marriage or genetically modified organisms, people have linked the natural and moral orders—and disorders. The stately rounds of the stars modeled the good life for Stoic sages; the rights of man were underwritten by the laws of nature in revolutionary France and in the newborn United States; recent avalanches in the Swiss Alps or hurricanes in the United States prompt headlines about "The Revenge of Nature." Nature has been invoked to emancipate, as the guarantor of human equality, and to enslave, as the foundation of racism. Nature's authority has been enlisted by reactionaries and by revolutionaries, by the devout and secular alike. In various and dispersed traditions, nature has been upheld as the pattern of all values: the Good, the True, and the Beautiful.[3]

For centuries, philosophers have insisted that there are no values in nature. Nature simply is; it takes a human act of imposition or projection to transmute that "is" into an "ought." On this view, we can draw no legitimate inference from how things happen to be to how things should be, from the facts of the natural order to the values of the moral order. To try to draw such inferences is to commit what has come to be called the "naturalistic fallacy"[4]—a kind of covert smuggling operation in which cultural values are transferred to nature, and nature's authority is then called upon to buttress those very same values. Friedrich Engels described this strategy in his critique of Social Darwinism, which he claimed was simply a reimportation back into the social realm of the Malthusian doctrines that Darwin had originally exported into the natural realm.[5] Engels's example shows that this sort of value-trafficking often has political consequences, as when medieval rulers defended the subordination of bulk of the population to the aristocracy and clergy on the grounds that it was as natural as for the hands and feet to serve the head and heart of the "body politic," or when early twentieth-century opponents of higher education for women argued that the natural vocation of all women was to be wives and mothers. Subordination and domesticity were thereby "naturalized": in such cases, contingent (and controversial) social arrangements were shored up by the necessity and/or desirability of allegedly natural arrangements. With examples like these in mind, some critics of alleged moral echoes of natural orders, such as the nineteenth-century British philosopher John Stuart

Mill, have condemned the naturalistic fallacy as not just logically false but morally pernicious to boot: "Either it is right that we should kill because nature kills; torture because nature tortures; ruin and devastate because nature does the like; or we ought not to consider at all what nature does, but do what it is good to do."[6]

Why, then, does the moral resonance of nature persist so stubbornly? Critical thinkers have spilled oceans of ink in attempts to pry "is" and "ought" apart. Despite their best efforts, however, the temptation to extract norms from nature seems to be enduring and irresistible. The very word "norm" epitomizes the mingling of the descriptive and prescriptive: it means both what usually happens and what should happen: "Normally, the cranes migrate before the first snow." I am under no illusion that another attempt to put "is" and "ought" asunder will succeed where the likes of Hume, Kant, Mill, and many other luminaries have failed. Rather, I want to understand *why* they have failed: why, in the teeth of such sterling counsel to the contrary, do we continue to seek values in nature?

I do not think the answer to this question lies just in an account of popular error, vestigial religious beliefs, or sloppy habits of thought. This is a case not of simple mass irrationality but rather of a very human form of rationality—and hence the subject of a philosophical anthropology. My line of inquiry will be to excavate the sources of the intuitions that propel the search for values in nature. In various times and places, these intuitions have expressed themselves in the most luxuriantly diverse forms—as diverse as the

eternity in our hearts, desire for understanding

efflorescence of nature and culture themselves. But the core intuitions underlying all this diversity of norms grounded in natures have something in common. At their heart is the perception of order—as fact and as ideal.

Some examples of the different ways natural and moral orders have been intertwined will help make the problem vivid. Because nature is so rich in orders, the analogy between natural and human orders can take many forms. Over the millennia, the authority of nature has been enlisted in support of many causes: to justify and to condemn slavery, to praise breastfeeding and to blame masturbation, to elevate the aesthetic of the sublime over the beautiful, and to undergird ethics by appeal to instinct or evolution. It would take many volumes (yet to be written) to do justice to this long and motley history and just as many volumes to describe the diverse natural orders used to represent and often legitimate these diverse norms. But certain forms of order recur over and over again, from Greco-Roman antiquity to yesterday's newspaper. At least within the Western intellectual tradition (the only one I am even partially qualified to write about), three in particular have exerted strong and lasting influence on both learned reflections and popular intuitions: specific natures, local natures, and universal natural laws.

2 SPECIFIC NATURES

Like all truly interesting words, "nature" is a *mille-feuille* of meanings. It can refer to everything in the universe (sometimes including and sometimes excluding human beings), to what is inborn rather than cultivated, to the wild rather than the civilized, to raw materials as opposed to refined products, to the spontaneous as opposed to the sophisticated, to what is native rather than foreign, to the material world without divinity, to a fruitful goddess, and a great deal else, depending on epoch and context. Scholars and etymologists compile long lists of such meanings and trace their tangled histories.[1] It would be folly to attempt to single out just one of these many definitions and nuances as primordial; the utility of the concept depends on the complexity of the word. But there is nonetheless one sense of "nature" in at least most of the major European languages that is at once ancient, enduring, and powerful: nature as the essence of a thing, that which makes it what it is and not something else, its ontological identity card. Because the prototypical (but not the sole) examples are organic species, I will refer to this scarlet thread in the skein of "nature" definitions as *specific nature.*

Specific natures embrace the characteristic form of things, be they chestnut trees or copper or foxes, their properties (flowering, reddish, cunning), and their tendencies (to grow from seeds, to take a polish, to breed in the winter). Specific natures determine how a certain kind of thing—animal, vegetable, mineral—looks and behaves. It is possible to alter both appearance and conduct, but only by constraining or "doing violence to" specific nature: it is the nature of flames to leap upward, but they can be constrained by screens; it is the nature of strawberries to ripen in spring, but they can be forced to grow in winter greenhouses. Especially for organisms, specific natures plot a path of development: it is natural for a pine tree to grow upward and a baobab to grow outward, for a tadpole to mature into a frog and a blackberry bush to bear summer fruit, for a puppy to become a dog and a lamb to become a sheep. Specific natures define the dramatis personae and plots of the universe.

Specific nature is the oldest and primary meaning of the ancient Greek word *physis* and the Latin word *natura*, both of which are etymologically linked to generation and growth. *Physis* shares its root (so to speak) with the word for "plant" and suggests self-directed growth and the species form that results.[2] *Natura* is formed from the verb "to be born"; we also catch the echo of this etymology in the words "nativity" and "innate." Greek and Latin meanings of *physis* and *natura* were intertwined already in antiquity; almost all modern European languages in turn derive their words for "nature" from the Latin *natura* and took over the meaning of specific natures lock, stock, and barrel.[3] Even cultures and languages with very different lineages and notions of the world make use of kindred concepts. The

Sanskrit word *dharma* is anchored in "the principle of order, regardless of what that order actually is" and includes specific natures: "The nature of an individual is the source of his own dharma and that of the group to which he belongs; it is the nature of snakes to bite, of demons to deceive, of gods to give, of sages to control their senses, and so it is their dharma to do so."[4] Ethnobiologists testify to the prevalence of not only classifications of plants and animals into species-like groups (sometimes designated as "folk" or "generic species") but also to the "commonsense assumption that each generic species has an underlying causal nature, or essence, which is uniquely responsible for the typical appearance, behavior, and ecological preferences of the kind."[5]

The metaphysics of specific natures is both mysterious, in the sense of not being accessible to easy scrutiny, and historically mutable: ancient Greeks imagined it as an inner principle; moderns conceive of it as DNA, or chemical composition, or a set of rules analogous to a computer program. More generally, specific natures refer to traits that are inborn or spontaneous, as opposed to those imposed by art or education. They can coexist with a variety of cosmogonies and theories about the world. Specific natures are compatible with an Aristotelian teleology that understands things as striving toward an appointed end (the rock striving to rest at the center of the Earth, the acorn striving to become a full-grown oak) or with materialist explanations that appeal to DNA and biochemical pathways. Specific natures may be implanted at the creation by divine decree or simply exist for all eternity or (as in evolutionary theory) come into being and pass away. But whatever the science and metaphysics invoked to ground specific natures, the idea itself is resilient. It has been attacked by nominalists in philosophy and by Darwinists in science, and contradicted by technologies as old as grafting and as new as genetic engineering. It has been deplored as the basis of racism and ridiculed as a doomed attempt to divvy up experience into tidy parcels. Yet species nature as idea and practice persists, deep-rooted and widespread.

The practice that goes hand in glove with the idea of species nature is classification, and at the most general level the very existence of common nouns testifies to the human cognitive necessity of lumping things into categories. We can barely conceive of a world in which every single thing is an irreducible individual, each so idiosyncratic as to be incommensurable with any other thing: a world of only proper nouns. Some scheme of categories is a precondition for both language and experience. But species natures are not just any categories. They carry nestled within themselves the essence and the narrative of their being: not only what they are at the moment but also what they have been and will become. All cultures concoct categories of convenience, connoisseurship, and caprice: household appliances, weeds, symphonies. A Borgesian imagination could make a game of this: the category of all clouds visible within an hour of dawn, toys broken by children who go without naps, urban winds that rush between tall buildings and out of subway stations. No one would mistake these categories, humdrum or fantastical, for specific natures. This is not simply a matter of manufacture: household appliances may be manmade, but weeds are not—although both categories are forged by convention. We can carve out categories of naturalia (clouds, plants, winds) as well as artificialia (washing machines, orchestral compositions). These categories may be useful, even indispensable for the business of everyday life: money is not a specific nature, but it makes the world go 'round. What is missing, however, is the integrity of a specific nature, which solidifies its properties and history into a stable kind, with a recognizable gestalt and predictable tendencies. In the organic realm this integrity is most often expressed in the process of faithful reproduction: like reproduces like.

The order of specific natures is typically disrupted by generation gone awry: monsters that transgress species boundaries or, especially in the Christian tradition, forms of sexuality that do not aim at reproduction, including homosexuality. The order of specific natures has been used to support an ideal of authenticity and defame an equally long-lived specter of the unnatural, most recently in debates about homosexual marriage. Aristotle returns to the criterion that true specific natures reproduce faithfully whenever he seeks to distinguish genuine natures from the products of chance, art, or convention. He counters the arguments that the universe might have come about by chance rather than through the regularities of specific natures with the example of reproduction: "From this it is evident that something of the kind really exists—that, in fact, which we call

[margin handwritten note: human nature fallen (vc intended)]

'Nature,' because in fact we do not find any chance creature being formed from a particular seed, but A comes from a, and B from b; nor does any chance seed come from any chance individual."[6] What is the difference between a genuine specific nature like human beings and artifacts like beds? Humans propagate other humans, but big beds do not beget little beds.[7] It is "contrary to nature [*para physin*]" for money to bear interest, because money does not possess a specific nature and therefore cannot reproduce itself as parents bear offspring.[8] In this last example, the normative and the natural interlock (in ways that will be consequential for centuries' worth of legislation on usury): obviously, money in Aristotle's Athens *was* lent at interest; hence his complaint. In this case, the unnatural is not the impossible but the undesirable.

Metaphorical violations of the prerogatives of specific natures to reproduce are bad enough; literal violations are still worse in Aristotle's book. Ideally, specific natures reproduce themselves and thereby the order of the world; even slight deviations from perfect copies are deemed monstrous: "Anyone who does not take after his parents is really in a way a monstrosity, since in these cases Nature has in a way strayed from the generic type."[9] For Aristotle, monstrosity is a continuum that begins when the offspring fails to replicate its male parent (in this sense, all daughters partake of deformity) and stretches to the extreme point when it does not even resemble its parents' species. The details of Aristotle's theory of generation need not concern us here, only the steely link that binds species nature to reproduction and therefore monstrosity—reproduction gone awry—to the subversion of species nature (fig. 2). This is why Christian theologians in the Aristotelian tradition such as Thomas Aquinas considered bestiality a worse sin than adultery or other sexual transgressions: such pairings overstep a boundary between species natures drawn by the "author of nature."[10]

Specific natures guarantee an order of things. Aristotle brandishes them as his weapons whenever he fights off philosophical opponents who claim the universe is the result of mere chance; they are still the sharpest arrows in the quiver of those who assert the workings of design over the meanderings of evolution. But it is not necessary to invoke the guiding agency of a deity or to personify nature (Aristotle's nature does not deliberate)[11] or even to sum all

Figure 2
Man-pig. Ambroise Paré, *Les Monstres et les prodigies* (1573).

the specific natures into one cosmic, all-embracing Nature in order to appreciate the regularity ensured by specific natures. Nor need one subscribe to the full Aristotelian view about specific natures. As Kant recognized, the stability of specific natures is a precondition for experience, over and above the psychological laws of association: "If cinnabar were sometimes red, sometimes black, sometimes light, sometimes heavy, if a man changed sometimes into this and sometimes into that animal form, ... my empirical imagination would never find opportunity when representing red color to bring to mind heavy cinnabar."[12] We can barely imagine a world without specific natures, in which everything would constantly be morphing into everything else and what a thing is would be no guide to what it was and will be. Nevertheless, the order of specific natures is not simply the ground zero of all orders; it is a distinctive order in its own right and, as subsequent chapters will make clear, not without alternatives.

3 LOCAL NATURES

Local natures are about the power of place. They refer to the characteristic combinations of flora and fauna, climate and geology that give a landscape its physiognomy: the desert oasis or the tropical rain forest, the Mediterranean shore or the Swiss Alps. The modern science of ecology studies the way organisms and topography interweave to create distinctive local natures, but long before there was such a science, people registered the order of local natures as the familiarity of home or the strangeness of the exotic. Since ancient times, local natures have been regarded as tightly interwoven with local customs. When the Greek historian and traveler Herodotus visited Egypt in the fourth century BCE, he described how, by back-home standards, both nature and custom were turned upside down, with the Nile flowing south to north and men and women reversing roles: "Just as the climate that the Egyptians have is entirely their own and different from anyone else's, and their river [the Nile] has a nature quite different from other rivers, so, in fact, the most of what they have made their habits and their customs are the exact opposite of other folks'. Among them the women run the market and shops, while the men, indoors, weave; ... The women piss standing upright, but the men do it squatting."[1] Local natures exhibit the same sort of regularities that local customs do: by definition, they are neither

neither uniform nor universal, but they nonetheless are predictable, within geographic limits. Seen globally, local natures make for a patchwork quilt of fields and forests, tropics and tundra. But within each patch, inhabitants mostly know what to expect, most of the time. This is the order of nature's customs, to which human customs are closely attuned. Calendars such as the gorgeously illuminated *Très Riches Heures* (*The Very Rich Hours*, a book of prayers keyed to the ecclesiastical year) of the Duke de Berry (created 1412–1416) illustrate the intertwining of the seasons with the cycle of human activities, as do countless proverbs and poems.

Although the interplay of local nature and custom lives on not only in poetry but also in whole schools of geography and *Annales*-style history, the metaphysics of custom that united both within a common framework from antiquity through the Enlightenment has been submerged, though never entirely lost. The salient aspects of this doctrine were: first, the distinctiveness of local nature and customs; second, the concord between them; third, the plasticity of both, often modified in concert; and fourth, a model of interaction that was integrative rather than mutually exclusive. Its locus classicus was the ancient Greek Hippocratic text, *Airs Waters Places* (5th c. BCE), in which itinerant physicians were advised on how to treat the inhabitants of various topographies and climates; its best-known political expression was the Baron de Montesquieu's Enlightenment treatise *On the Spirit of the Laws* (1748), which posited a harmony between peoples, climates, topographies, and laws. An updated version of local natures became the nucleus for a vast nineteenth-century scientific research program inspired by the Prussian naturalist Alexander von Humboldt's *Views of Nature* (1807), which classified landscapes into "physiognomies."

In *Airs Waters Places*, traveling doctors are enjoined to study the effects of winds, seasons, stellar aspects, water, soil, and mode of life on the inhabitants of different places in order to diagnose and treat typical local diseases. Medical knowledge gleaned from one locale cannot be readily generalized to another, unless they resemble one another along these crucial dimensions. Although specific natures are uniform in time and space, their complex combination and mutual modification produce distinctive local gestalts. Asia differs from Europe, for example, "in the nature of all its inhabitants and of all its vegetation. For everything in Asia grows to far greater beauty and size; the one region is less wild than the other, the character of the inhabitants is milder and more gentle."[2] Herodotus also draws attention to the natural specificity of different locales, and to providential compensations among and within them. Gold may be more plentiful in India than in Greece, but Greece is blessed with a more temperate climate; Arabia abounds with poisonous snakes, but since the female of these species bites off the head of the male during mating, and is in turn killed by her offspring to revenge their father (a kind of serpentine *Oresteia*), their numbers are kept in bounds.[3] Although the Hippocratic authors invoke neither the gods nor providence, a similar compensatory and complementary logic can be detected in some descriptions: whereas Asian peoples and cattle are typically more well-grown and handsome than their European counterparts, courage and diligence cannot flourish under Asian conditions.

Local natures and local customs work in tandem. The Asians lack spirit, endurance, and industry both because their climate is mild and equable *and* because they are ruled by despots, who offer no incentive to take military risks or work hard.[4] A similar logic governs the Hippocratic account of the tribe of Longheads, who stretch the heads of their young children with bandages and all manner of other devices because the shape is considered attractive. This custom was eventually supplemented by nature, *physis* and *nomos* cooperating with one another: "Custom originally so acted that through force such a nature came into being; but as time went on the process became natural, so that custom no longer exercised compulsion."[5]

The idea of local natures is ancient, but it was given new momentum in the seventeenth and eighteenth centuries by elaborate theories in both natural history and natural theology (which often went hand in hand) about the equilibrium of natural systems, from the solar system to the ecosystem. Nature in its entirety was reimagined as more coherent: a harmonious whole whose interlocking parts were in delicate equilibrium with one another. The eighteenth-century Swedish naturalist Carolus Linnaeus described this system as an "oeconomy"; the modern term is "ecology."[6] Both words derive from the ancient Greek word for household (*oikos*), a self-contained unit dependent on a division of labor and continuous give-and-take among its elements. The equilibrium of the *oikos* is dynamic because it is achieved through tension, a rope made taut and straight by a tug-of-war between opposing forces. Its harmony emerges only at the macroscopic level, both in space and time; viewed microscopically, at a pinpoint place and moment, the members of the household—or the organisms of an ecology—incessantly jostle, elbow, and adjust to one another. The idea of the *oikos* is ancient, as is that of what would now be described as a local ecology. Long before there was a science of that name, people remarked upon the order of local natures: the characteristic combination of flora and fauna, climate and geology that gives a landscape its recognizable physiognomy, whether Death Valley or Brazilian rainforest, Mediterranean shore or Siberian tundra. Key to all of these local ecologies was the idea that the elements of each formed a harmonious (and sometimes precarious) whole, poised in delicate equilibrium.

If monsters are the prototypical disruption of the order of specific natures, disequilibria play the same role for local natures. This was and remains especially the case when human activity seems partly or wholly responsible for upsetting nature's balance. The nature that takes its revenge in today's headlines about disasters is neither personified nor deified. Mother Nature may be caricatured in cartoons, but with an irony that undercuts any serious construal of intentions in hurricanes or wildfires. Even proponents of the Gaia hypothesis, which envisions the Earth as a living organism, quickly admit that it is not "alive in a sentient way, or even alive like an animal or bacterium."[7] Rather, vengeful nature is a self-regulating system, like a thermostat or the governor of a Watt steam engine.[8] The general principle is the same, whether the system is an organism, the solar system, or a machine: small shocks will push it out of equilibrium, but only temporarily; when the system is jolted too violently and for too long, however, it starts to fluctuate chaotically. Organs fail, moons spin off on tangents, machinery explodes. The general mechanism that triggers nature's revenge is just that: a mechanism.

But why revenge? Why not just unintentionally caused disequilibria—negligent, perhaps, but not malevolent? Why does guilt still saturate our understanding of natural disasters, however it is channeled? Nature's revenge is never invoked unless there is human complicity in the disaster, even if no one deliberately willed the devastation: the greedy developers who sold flood-plain real estate; the lazy politicians who didn't maintain the dikes; the corrupt officials who allowed construction companies to ignore building codes in an earthquake zone; the self-indulgence of industrialized nations that refuse to reduce carbon emissions. It is telling that the phrase "nature's revenge" cropped up in editorializing on the March 2011 disasters in Japan almost exclusively in connection with the accident at the Fukushima nuclear power plant—not the earthquake or tsunami that triggered the accident, even though the earthquake and tsunami cost three orders of magnitude more lives and were demonstrably more natural. Only where human hubris, greed, or sloth can be laid bare does "nature's revenge" come into play, no matter how horrendous the havoc wrought by other, unequivocally natural disasters.

The paradoxical consequence of ever-deeper inquiry into the natural causes of devastation has been to expose human motives, turning natural disasters into sagas of crime and punishment—so far, mostly at the local level, though this may be changing in debates over global climate change. Local natures are thrown out of joint not by monsters but by disequilibria: upset the delicate balance of the elements, and the whole is threatened with disaster—with nature's revenge. Like the order of specific natures, the order of local natures has a long history, from Hippocratic medicine to contemporary worries about genetically modified organisms.

4 UNIVERSAL NATURAL LAWS

Universal natural laws, in contrast to nature's local customs, admit of no exceptions—at least not by mortals. They define a uniform and inviolable order, everywhere and always the same, exhibiting ironclad regularities. If the science of specific natures is taxonomy and that of local natures is ecology, the science of universal natural laws is celestial mechanics. The inexorable progression of the stars and planets through the heavens provides the model for a perfectly regular world of changeless change. In the next room and in the remotest galaxy, the same natural laws hold. The violation of a universal natural law is a miracle—or pure randomness. The prototype for the order of universal natural law is universal gravitation, set forth in all its magisterial generality by Isaac Newton in his *Mathematical Principles of Natural Philosophy* (1687). However, like specific and local natures, the idea of universal natural laws stretches back to antiquity, especially in the context of mathematical sciences such as astronomy and optics. Seneca appealed to a possible "law" of comets, and Pliny called for a "law" of the maximum elongation of Venus and Mercury.[1] The doctrine of universal determinism also has an ancient lineage, but at least in the Latin West, its ancestral home was not natural philosophy but Christian theology, in various controversial doctrines of predestination, from Augustine through Calvin.

Until the seventeenth century, however, the terminology of laws applied piecemeal to particular regularities (especially in astronomy) rather than to nature as a whole. Other terms, such as "rules" (*regulae*) or "axiom" (*axiomata*), were at least as widespread.

Aristotelian natural philosophy and its medieval Latin interpreters understood *scientia* worthy of the name to be certain, causal knowledge of universals—but not *universal* knowledge of universals. Some disciplines might be certain but not causal: the so-called mixed mathematical disciplines of premodern astronomy, optics, and harmonics offered impressive mathematical models of the movements of heavenly bodies or the reflection and refraction of light but without reference to physical causes. Other disciplines might be too mired in particulars to permit generalizations: practical medicine, which must deal with the diversity of individual bodily complexions, belonged to this category. And still other phenomena were either too rare (e.g., an aurora borealis or a rain of blood) or too opaque to the senses (e.g., the workings of a magnet or a poison) to reveal anything more than observed correlations. The certainty of scholastic natural philosophy was bought at the cost of circumscribing its range: it offered explanations of what happened always or most of the time (in Aristotle's oft-repeated phrase), not of phenomena that occurred once in a blue moon or varied from individual to individual or thwarted causal inquiry altogether—even if all of these events were considered to be perfectly natural. In many fields of useful knowledge, such as practical medicine or engineering, it was taken for granted that there would be exceptions to whatever generalizations held for the most part.

There was no standardized medieval and Renaissance vocabulary for such imperfect regularities: in addition to "rules" (*regulae*), words such as "precepts" (*praecepta*), "axioms" (*axiomata*), "aphorisms" (*aphorismi*), "customs" (*consuetudines*), and (especially in the context of astronomy and grammar) "laws" (*leges*) might be used, depending on discipline and context.[2] Regardless of the terminology, what was *not* meant was an inviolable and immutable natural law in the modern sense, valid everywhere and always. Instead, nature was understood as a patchwork of regularities of different kinds, jurisdictions, and degrees of strictness. Some regularities were mathematical, but applied to a restricted range of phenomena (those in which form preponderated over matter, as in the case of light, or in which the matter was of a special kind, as in the case of the celestial aether); others derived from the specific nature of the substance (e.g., fire burns everywhere); still others were observations coalesced from long experience (e.g., that certain kinds of clouds portend fair weather). Nature was orderly but not exact, following customs that occasionally admitted of exceptions rather than strict laws.

The concept of natural laws—uniform, universal, and inviolable—emerged in the course of the seventeenth century from a tangled skein of theology, natural philosophy, and mixed mathematics.[3] Decisive for the ascent of the idea of nature governed by laws was a voluntarist theology that imagined God as a "divine legislator" who imposed "laws" upon nature as an absolute monarch imposed laws upon a kingdom. Leading figures in the scientific revolution, including René Descartes, Robert Boyle, and Isaac Newton, all subscribed to some version of this view of natural law as the expression of divine free will. This legal metaphor at first caused some perplexity even among its adherents: How could unthinking matter "obey" laws that required conscious assent? How could the idea of laws, which were notoriously subject to violations and local exceptions, capture the sense of a divine edict valid everywhere and always? And how could such inviolable natural laws be reconciled with the religious doctrines of biblical miracles?[4]

Machine analogies, especially to clockwork, played a central role in establishing the idea of universal, inviolable natural laws despite these difficulties. The writings of the seventeenth-century English natural philosopher Robert Boyle, who pondered the problems of applying the idea of law to all of nature, offer some striking examples. Most of Boyle's examples of how the macroscopic properties of phenomena could be explained by appeal to the hypothetical shape, size, and number of microscopic corpuscles could have (and sometimes did) come from Lucretius's Epicurean poem *On Nature* (1st c. BCE): for example, the reason honey is sticky is because its constituent particles have little hooks, like brambles. However, there was one kind of contemporary machine to which Boyle repeatedly returned when in search of analogies and metaphors to illuminate the mechanical philosophy: clockwork—and not just any clockwork, but the fanciest clocks known to early modern Europeans, tourist attractions such as the astronomical clock in the Strasbourg cathedral, constructed 1570–1574 by a team of Swiss artisans led by the mathematician Conrad Dasypodius (fig. 3).[5]

What fascinated Boyle about the Strasbourg clock, with its multiple dials to track the motions of heavenly bodies and elaborate displays of crowing cocks, multi-tune carillon, and parading automata, was the way in which the artificers had built the future into its workings. Once set in motion, all the processes unfolded without any further intervention on the part of the clock's makers. Nature was just such an engine, Boyle argued, built by God:

> [I]t is like a rare clock, such as may be that at Strasburgh, where all things are so skillfully contrived, that the engine being once set a moving, all things proceed, according to the artificer's first design, and the motions of the little statues, that at such hours performs these or those things, do not require, like those of puppets, the peculiar interposing of the artificer, or any intelligent agent employed by him, but perform their functions upon particular occasions, by virtue of the general and primitive contrivance of the whole image.[6]

Even those phenomena that strike human observers as anomalous—earthquakes, volcanic eruptions, new stars, and the other oddities excluded from Aristotelian natural philosophy—had been foreseen by the Creator from the outset and built into the divine clockwork. God might, Boyle admitted, very occasionally overrule His engine with a miracle, but Boyle argued that a genuine miracle could be

Figure 3
Astronomical Clock (1570–74), Strasbourg Cathedral. Photo courtesy
David Iliff. License CC-BY-SA 3.0. Available at https://commons.wikimedia
.org/wiki/File:Strasbourg_Cathedral_Astronomical_Clock_-_Diliff.jpg.

distinguished from the "predesigned" marvel by a suspension of the otherwise inviolable regularities of matter and motion.[7] Boyle's motives for likening nature to the Strasbourg clock were as much theological as natural philosophical. A clock, no matter how elaborate, had no agency of its own. Boyle objected to the personification of nature and warned that to grant creatures other than men and angels intelligence and rationality was to risk idolatry: had not both Jews and pagans lapsed into the worship of the sun and moon and other heavenly bodies?[8] Boyle's analogy between the entirety of nature, including anomalies, and what he called the "predesigned"—we might be tempted to say "preprogrammed"—workings of a real machine, the Strasbourg astronomical clock, captures the idea of inviolable, uniform, and universal "laws of nature," understood as more fundamental than mere empirical "rules."[9]

Like Boyle, and in opposition to Gottfried Wilhelm Leibniz, Newton insisted that natural laws could be suspended as well as imposed by God: in principle, miracles were still possible; God's free will was unfettered by any constraints whatsoever, including the natural laws He himself had ordained.[10] But in practice, such deviations were vanishingly rare. The divine engineer who foresaw all eventualities had no need to suspend the workings of the machine of nature. During the Enlightenment, Newtonian natural philosophy and its vision of a universe governed by universal laws, everywhere and always the same, fired the imagination of philosophers and political reformers, who sought equivalent universal laws for the human realm (fig. 4). Both the

Voulez-vous être heureux? écoutez la Nature.

Figure 4

Pierre-Platon Blanchard, "Do you want to be happy? Listen to Nature."
Catéchisme de la nature ou Religion et morale naturelles (1794).

Declaration of Independence (1776), in which the American colonists broke their ties to the British crown, and the *Declaration of the Rights of Man and the Citizen* (1789), proclaimed by the French revolutionary National Assembly, enlisted the language of rights guaranteed by nature and therefore universal and inalienable. Like the orders of specific and local natures, the order of universal natural laws is still present in today's imagined moral orders, for example the campaign for universal human rights that know no national borders or local jurisdictions.

The order of natural laws became a secular metaphysics during the Enlightenment, despite its origins in the theology of a completely free divine will that imposed—and in principle also occasionally revoked—its dictates on the entire universe. The only trace of the voluntarism that had initially inspired this order lay in its disorder: the exercise of will, either divine or human, unbound by the deterministic laws that ruled in every other realm. Acts of divine will produced miracles; acts of human will, moral freedom. These became the only loopholes in the otherwise universal jurisdiction of natural laws.

5 THE PASSIONS OF THE UNNATURAL

In the course of many centuries and cultures, each of these natural orders has been used to imagine and justify various moral orders. What is distinctive about the three I have singled out is that they have been long-lived, polyvalent, and evocative of powerful emotions when violated. The emotions in question are characteristic and vehement; they are also unusual among emotions in combining strong feeling with intellectual judgment: the passions of the unnatural. Each of these three natural orders has been used to define and oppose a distinctive form of the unnatural: the monsters that violate the order of specific natures; the imbalances that capsize the order of local natures; the indeterminism that breaks the order of natural laws. It is a striking fact that these versions of the unnatural also provoke distinctive emotional responses: horror, terror, and wonder, respectively. These are the emotions—or better, passions, in the original sense of the term as an extreme state that we suffer rather than merely feel—that register a breach of order. Although, for example, horror and wonder may seem poles apart as states of experience, they are linked by deep ties, as evidenced by the strange tendency of one passion to tip over into the other. Horror and terror are more obviously related to one another, but the peculiar terror evoked by "nature's revenge" also shows revealing affinities to wonder.

These passions form a triplet, united both by their inter-relationships and by a shared tendency to blur moral and natural stimuli. They are the subjective side of the objective perception of a disorder so dramatic that even nature quakes.

Wonder, horror, and terror are true passions, not mere emotions. In the root sense of the word "passion" (from the Greek *pathema*, the Latin *passio*; cf. the German *Leidenschaft*), passions are suffered like an illness (they have the same root as "patient"), things that befall rather than move us, not so much states as sieges of the soul. In contrast to the emotions, first conceived in the eighteenth century as movements in the nerves and brain, or the still more delicate sentiments and feelings, passions don't belong to us; we belong to them.

Whether the disorder that triggers one of these passions is natural or moral is often extremely difficult, if not impossible, to ascertain. Is, for example, the horror of monsters that apparently cross species, such as the hoax image of a human ear apparently growing from a mouse's body, a response to a trespass against a natural boundary or to a transgression against a moral taboo that prohibits bestiality or scientific hubris?[1] Is the terror unleashed by a flood or an avalanche simply the magnified fear of extreme danger to life and property, or is it the fear deepened by guilt over partial responsibility for the catastrophe?[2] Is the wonder of miracles (or free will) evoked by snapping the chain of material causation, or by asserting volition, human or divine, in deliberate defiance of all constraints (fig. 5)? Even to pose these questions as either/or choices seems strained: it is characteristic of these passions to blur the distinction between the moral and the natural.

These subjective responses to the unnatural, as variously defined by kinds of natural order, suggest that there is a cognitive component to at least some passions. Horror, terror, and wonder are triggered when a major disruption of order (whether moral or natural or both) is registered as such: an act of perception and judgment that presumes some acquaintance with the particular sort of orderliness that has been breached—that, as the phrase goes, "something is not right." One must, for example, know a fair amount about local climate and flora and fauna to discern that the swallows have not returned from their annual migration or that the monsoon rains are very late this year. Knowledge can reclassify an apparent miracle into a completely predictable event: as Thomas Aquinas noted, astronomers do not wonder at eclipses that dumbfound unlettered peasants.[3] But it is also the case that knowledge of nature's rules can enlarge awareness of possible exceptions: for the astronomer well acquainted with lunar perturbations, an inexplicable wobble, however slight, is a potential miracle. Wonder, terror, and horror are the not the only cognitive passions[4]—curiosity would be another candidate—but they are among the most powerful.

Despite dramatic differences in emotional texture, wonder, terror, and horror all contain a moment of astonished disbelief. They are the eye-rubbing passions of incredulity: "I can hardly believe my eyes" (fig. 6). Like the more familiar moral passions, such as anger ignited by injustice

Figure 5

Henri-Frédéric Schopin, *The Children of Israel Crossing the Red Sea* (ca. 1855). Courtesy Bridgeman Images.

"Is the wonder of miracles evoked by snapping the chain of material causation, or by asserting violation, human or divine, in deliberate defiance of all constraints?"

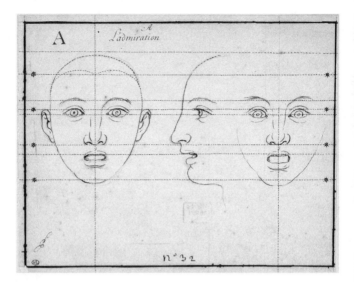

Figure 6
Charles Le Brun, *Wonder* (1667). The Louvre, Paris, France.

or grief undammed by loss, they are sudden, intense states that pounce upon us unawares. Literary scholar Philip Fisher has written brilliantly about these "vehement passions" and the way they flood over us, eclipsing everything but the object that evoked them and fusing the self into a momentary monolith: "Unlike the feelings, the affections, or the emotions, the passions are best described as thorough. They do not make up one part of a state of mind or a situation. Impassioned states seem to drive out every other form of attention or state of being."[5]

The passions of the unnatural share these qualities of suddenness and vehemence to which Fisher draws attention. But they are not entirely "thorough" in his sense, that is, driving out "irony and all forms of double conscious" for their duration.[6] To register simultaneously doubt ("Can this really be happening?") and doubt overcome ("It *is*— horribly, terrifyingly, or wondrously—really happening!") is a distinctive form of double consciousness that has little in common with the wry half-smile of irony or the cool detachment of deliberation or the split consciousness of self-observation. All the vehemence and monomania of the passions are preserved, and yet the self fissions into doubter and believer, briefly coexisting in a single consciousness. It is precisely this double state, however fleeting, that marks out the passions of the unnatural as cognitive: they flag an order at once ratified and destroyed. Doubt is strong because the order it in effect upholds is so strong. Disbelief is the left-handed compliment paid to disorder: how is this possible? What force is powerful enough to wrench nature out of joint? Depending on the kind of natural order that is violated, horror, terror, or wonder affirms that the impossible has indeed occurred. The passions of the unnatural jangle the soul with an almost unbearably dissonant chord.

The passions of the unnatural must be distinguished from passions that register some breach of the strictly moral order, most notably indignation and outrage. As the "rage" in "outrage" signals, both of these responses are at root forms of anger. And anger makes no sense unless the culprit is a person who can be made responsible for transgressing norms that the rest of the community respects. It is telling that when anger is directed at a nonhuman actor—for example, a deer that has destroyed a vegetable garden—fury briefly transforms the animal into a person to whom malevolent motives can be imputed by the victim. But indignation properly so-called is rarely directed at a being—animal, infant, madman, nature—that cannot be reasonably expected "to know better." To shake one's fist at the sky or to rail against the elements as King Lear does in Shakespeare's play is either to impute rational agency and moral responsibility to some deity or the weather or prima facie evidence of madness. The gardener's anger does not go so far as to substitute remonstration with the marauding deer for a reinforced fence. In contrast, responsible adults who affront some norm can expect an earful of reproach as well as punishment. Indignation is seldom willingly silent: it is a form of anger that demands to be vented in words, not just discharged in deeds. We confront the culprit with the enormity of the transgression and, in so doing, recall him or her to the fold of shared values and accountability. For the object of indignation not to show contrition—not to acknowledge in words and bodily posture the legitimate claims of the violated norms—is to invite redoubled reproaches. For all its sound and fury, indignation ultimately aims to reintegrate the culprit into the community by shouting out its values and exacting a shamefaced acceptance of them.

Horror, terror, and wonder are, in contrast, dumbstruck passions. There is no temptation to scold a monster or berate a drought or upbraid pure randomness. Only when these natural disorders shade into human culpability does outrage tinge the passions of the unnatural. If the monster is believed to be the issue of a sinful union, if the drought is believed to be due to greedy land grabbers, if the decoupling of cause and effect is believed to be divine or demonic intervention—only then can indignation be unleashed. Disruptions of the moral order in these cases are regarded as complicit in disruptions of the natural order. Yet in extreme cases, the distinction between responses is as blurred as that between orders. Horror can be evoked, for example, by human atrocities so enormous that we question the bare humanity—the specific nature—of the perpetrator. Heinous crimes knowingly committed can provoke outrage so overwhelming that observers are rendered speechless—as if the evil exceeded merely human bounds. In some languages, acts that flaunt entrenched norms are described as not just "wrong" but also as "inconceivable" or "unbelievable" or even "impossible." In English, an atrocious act often evokes the question: "What kind of person could do a thing like that?" The implication is that the perpetrator is not a person at all but a moral monster that has transgressed the bounds of humanity. These are of course exaggerations: the trespasses were obviously all too possible and their perpetrators all too human. Yet these hyperboles (and the emotional responses they express) point to a deeply felt connection between violations of the natural and the moral orders, which between them cover the territory of what does happen and what should happen.

The purpose of delving into the characteristic passions of the unnatural is twofold. First, their specificity and intensity help to sharpen the outlines of the different kinds of natural order they monitor, as well as to reveal how significant these orders are for lived human experience—so significant that we have specialized and gripping emotional responses to perceived disorders. The very existence of such passions is evidence of sustained and discerning human attention to the orders exemplified in nature. By exploring the passions of the unnatural and their interrelationships, we learn something about order per se, whether moral or natural. The kinship of the horror-terror-wonder trio hints at a parallel kinship among various kinds of order.

Second, the passions of the unnatural offer insights into the sources of some fundamental moral intuitions. Moral intuitions cannot do the work of moral reflection, and reflection sometimes amends or discards the impulses that flow from intuitions. Not all of our intuitions survive severe scrutiny, and there may be good grounds for ignoring at least some of the promptings of the passions of the unnatural. Before, however, our intuitions can be embraced or rejected, they must first be identified and analyzed. Moreover, without some kind of moral intuition it is very difficult if not impossible to nourish reflection and propel the will. Even Kant, the philosopher most wary of moral feeling, remarked that reason alone may not be sufficient to determine the will objectively: some subjective inclination is required to prod human beings to act as they know they ought to. Only in the perfectly good will of the holy would reason and inclination coincide.[7] For beings like us, with one foot in the world of sense and the other in the world of reason, intuitions are needed to galvanize if not dictate to the will. Just because our moral intuitions are necessary but not sufficient to guide right action, some understanding of their provenance and power is an essential prelude to reflection.

The very existence of the passions of the unnatural is evidence of sustained and discerning human attention to the orders exemplified in nature. But why are these orders moralized? Why do the passions of the unnatural straddle the boundary between the normative and the natural so ambiguously? How can norms be derived from natural order of any kind?

Of all the nightmares that bedevil the collective human imagination, that of chaos is the most terrifying. Human history is stained with orders that have been bloody, tyrannical, and ruthless, orders that suffocate like an iron vise. And many philosophers and scientists have judged the order of nature to be heartless, inexorable in its workings and indifferent to human joys and sorrows. Order itself can become a nightmare. But the horrors of excessive order pale beside those aroused by no order at all. Endless civil war is a greater calamity than the most oppressive dictatorship; a universe formless and lawless is the ground zero of all cosmogonies, whether it is a deity or natural law that is called upon to create a cosmos worthy of the name. A land in which no promise is kept, in which the sun may or may not rise on the morrow, in which the past is no guide to the future, is a no man's land.

This nightmarish thought experiment conflates natural and human orders in a way that will raise eyebrows: isn't this just another instance of the naturalistic fallacy, mixing up "is" and "ought"? Though few may doubt that both natural and social chaos are, each in its own way, horrific, many thoughtful readers will query the equation of the two—and still more the identification of the two orders chaos annihilates. It is an axiom of modern thought that nature and society are distinct realms. As the British zoologist Thomas Henry Huxley put it in his 1893 Romanes Lecture: "Let us understand, once for all, that the ethical progress of society depends, not on imitating the cosmic process, still less in running away from it, but in combating it."[1]

Despite the force and familiarity of these objections, I would like to reopen the question of norms from nature. My argument hinges on a distinction between the content of specific norms—for example, those that prohibit stealing or lying as wrong—and a more general claim to what philosophers call *normativity*:[2] roughly, the justification that gives any and all norms their force. It is a notorious fact that specific norms vary dramatically across cultures and over time. This also holds for norms that invoke nature, which run the political gamut from apartheid-style racism to Green Party environmentalism. But normativity is a far more uniform and durable phenomenon: there is no known human culture, past or present, without any norms at all. The cross-cultural diversity of norms that is often cited as evidence for the relativity of all norms might equally well serve as evidence for the universality of normativity. A culture without norms is as much an oxymoron as nature without regularities. Pockets of anarchy and randomness are a far cry from total chaos.

Normativity is one of those bloated abstractions that make the mind go blank and eyes glaze over. But the meaning of normativity is quite simple: it is the quality of telling us what *should* be, as opposed to describing how things actually are. There are many houses in the mansion of "should," including the "shoulds" of how we should act, how we should know, and what we should admire—otherwise known as the Good, the True, and the Beautiful. What all these "shoulds" have in common is a certain wistful, counterfactual mood, a kind of subjunctive yearning: "If only things were the way they should be!" Normativity is the roof over the mansion of "should," the quality that allows us to recognize intellectually that there is a gap between what actually is and what should be the state of affairs—and moreover to experience regret at this mismatch.

The intensity of that regret can range from mild to intense, from an inward sigh over boorish conduct to a flash of anger at injustice. Exactly what triggers such responses is as varied as human culture and history. In some cultures, slavery was taken for granted but the sight of a woman's exposed ankle was shocking; in others, equality before the law was hallowed but extreme economic inequality accepted without protest. These divergences in specific norms can notoriously flare up into mutual outrage when cultures collide. But human beings who never experience indignation or outrage at *anything* are barely imaginable. Even in fiction, readers dismiss villains who bow to no "should" as cardboard characters, entertaining perhaps but unreal. To plumb the sources of normativity—what gives any and all norms their claim to authority over our judgment, if not always over our conduct—is a very deep, perhaps bottomless philosophical problem. For my purposes, it is sufficient simply to register the empirical fact that part of what it means to be human is to acknowledge some norm or another, to understand the force of "should," and to feel a stab of regret at the distance between what is and what should be.

All well and good, you may say, but what does the bare existence of some kind of norms have to do with order, much less with nature? Perhaps some cultural norms can be expressed as a coherent system, on the model of an idealized textbook version of Roman law. But, you will persist, most norms evolve slowly, in specific historical circumstances, and accrete like the layers of buried objects on an archaeological site rather than coalescing into some organic whole. Some norms may derive from ancient customs; others may have been introduced with new religions; still others result from concerted deliberation and debate. Over generations and centuries, even the most traditional societies refine old norms and adopt new ones. Although blatant contradictions may be weeded out, it seems highly unlikely that the outcome of such processes would be a tidy order. And when order among norms does exceptionally emerge, it is through the systematizing efforts of legislators, theologians, jurists, and philosophers, not through any intervention on nature's part. So why invoke order, much less natural order?

This is a plausible account of the development of specific norms, but it neglects the preconditions for norms in general – for normativity. Without a background of order, no norm can take hold. The very idea of a norm implies some consistency and generality—though not necessarily complete uniformity and universality. Norms are not ad hoc rules, improvised for the occasion, although it may take considerable reflection and ingenuity to apply them to the myriad of particular circumstances that might arise, as judges who must interpret past law and precedents in light of present cases know all too well. For much the same reasons that there cannot exist a purely private language, there cannot exist purely private norms: norms imply a community, which may be defined as narrowly as the inhabitants of a single village or as widely as all rational beings, but never contracted to a single individual. Moreover, norms imply a temporal horizon that stretches at least some way into the past and, still more important, into the future. Just how far in past and future directions depends on the reach of communal memory and expectations, both of which can be extended by cultural technologies, ranging from writing to life insurance. But no norm can be confined to the pinpoint present and remain a genuine norm. There must exist enough order to guarantee that norms that hold for my peers (however defined) will also hold for me and that today's norm will also hold tomorrow.

To return to our question: what do norms have to do with order, much less with nature? The answer is that if minimal conditions of order are not met, the idea of normativity crumbles: not just this or that specific norm, but any conceivable norm. Let us return for a moment to the nightmare of chaos. A situation that is so volatile and uncertain that what happened yesterday is no guide for today and today is no guide for tomorrow can support neither promises nor predictions. Where I walked yesterday with complete insouciance has today become a gauntlet to run; the

sin//holiness

neighbor who is now my friend may at any moment turn foe and then back again; the seasonal rains that sustain crops may or may not come. No one and nothing can be relied upon. Note that this is an anarchy more extreme than a Hobbesian state of nature: even in the war of all against all, self-interest makes one's rivals calculable. Strategic war games assume a rationality of self-preservation even among implacable enemies. But even this barebones order based on egotistical reckoning disappears in genuine chaos. Under such circumstances of complete uncertainty, even the crudest norms of reciprocity and revenge erode: I'll-scratch-your-back-if-you-scratch-mine and tit-for-tat assume some kind of temporal reach, a future into which it makes sense to extrapolate intentions. The minatory "should" assumes that the future tense "shall" has meaning. Normativity itself has no traction without some kind of order.

The connection between normativity and order runs deeper still. Not only is some minimal order the practical precondition for any kind of norms; normativity itself posits an ideal order. It takes a considerable effort of reflection to make such ideal orders explicit. From Hesiod and the Laws of Manu to the United Nations Charter and John Rawls's *A Theory of Justice* (1971), great works of literature, theology, and philosophy have envisioned orders in which to lodge particular norms of neighborliness, justice, filial piety, and human dignity. These orders are rarely rigorously systematic. They are less like mathematical demonstrations and more like architectures, in which diverse elements are structurally and stylistically combined into harmonious but contingent wholes, be they gothic cathedrals or high modern skyscrapers. The orders of norms are moreover as far from miscellaneous lists as they are from Euclid's *Elements*. When we do encounter list-like norms, for example, the dos and don'ts of the ancient Pythagoreans that survive in fragmentary texts, we are puzzled and immediately set about trying to reconstruct the missing connective tissue that would convert the apparent hodge-podge into some semblance of order: what could directives such as "Abstain from beans" and "Never step over a cross-bar" have to do with one another?[3] How do the parts of the moral edifice fit together and mutually support each other? Some degree of coherence is particularly important as a guide to moral reflection when new, previously unimagined quandaries emerge, whether in the context of advances in reproductive technologies or animal welfare. Recurring to the architectural metaphor, the style and structure of the original building guide the design of the new extension. Normativity presupposes order, both practically and theoretically.

But, you will persist, why drag nature into this? Even if normativity requires some kind of order, don't human societies generate plenty of order—spontaneously, inventively, incessantly? Why would connecting manmade orders to nature lend them one whit more authority or solidity? Why pretend that human norms echo nature? Perhaps human nature has something to do with this creative craving for orders by which to live, just as we are a species that builds homes to inhabit, but why is Nature writ large relevant to these activities? Isn't this just the residue of a religious argument that sees in nature God's creation and proxy, so that the alleged authority of nature's order is just divine authority at one remove?

These objections and suspicions are not without grounds. It is indeed puzzling why a putative parallel between natural and moral orders should do anything except multiply entities: why isn't one order—the human order, made by and for humans—sufficient? Appeals to natural order are at least more comprehensible, if not more philosophically defensible, if nature is understood to be itself divine or a divine creation and therefore as mirroring divine will. This is an idea found in many religious traditions and not just in the monotheistic Abrahamic religions of Judaism, Christianity, and Islam, which all accept some version of a creation story.[4] Aristotle in contrast believed in the eternity of the world, uncreated and unending. Yet even he remarked that "both barbarians and Hellenes, as many as believe in gods" thought that there was something divine about the heavens, the starry realm above the orbit of the moon, perfect and unchanging—"that immortal is closely linked with immortal."[5] But these are all appeals to belief, not to argument. Skeptics are right to insist upon a human justification—or at least explanation—for human practices.

That explanation lies in an equally human trait: the irrepressible urge to represent, to make the invisible visible, to render immaterial ideas concrete and tangible. Philosopher Ian Hacking asserts this trait as the basis of a philosophical anthropology: "*Human beings are representers*. Not *homo faber,* I say, but *homo depictor*. People make representations."[6] To draw geometric diagrams, to paint and sculpt, to construct models of the cosmos, to imagine allegories of abstractions, to fashion symbols and images of all kinds—these are all expressions of this fertile propensity to represent. I second Hacking's insight (advanced in the context of debates over scientific realism) and would like to extend it to the strange proliferation of what would seem to be gratuitous analogies between natural and moral orders. It is an empirical fact that humans *do* use natural orders to represent moral orders. They do so even when the natural order possesses no more or less authority than the moral order it purports to model; they do so when it is possible to find alternative models—in for example art or mathematics or technology. Why?

7 THE PLENITUDE OF ORDERS

Nature possesses at least two advantages over all other candidates for such models. First, it is everywhere and always on display, available and familiar. It requires no creative labor to fabricate these models, although it takes great ingenuity to craft persuasive analogies between, for example, beehives and monarchies or between the inexorable march of the sun along the ecliptic and the justice of the law courts. A serviceable model must be thing-like: an object that can be contemplated publicly as well as privately, accessible to the senses as well as the intellect, and possessed of sharp-edged solidity, the sort of thing you can stub your toe against. Whereas social orders are notoriously difficult to locate ("Where is society?" as British Prime Minister Margaret Thatcher once notoriously asked) and to study in their detailed workings (all that obscure talk about the "invisible hand" and "collective consciousness"), the natural orders used to model them are as obvious as rocks—though often just as opaque in their causes and composition. This is one reason why for centuries, beehives and anthills have been used to model human societies; more recently, economists have appealed to hydraulics to represent supply and demand in the market (fig. 7). storehouse

Second, and more important, nature is the repository of all orders. Whether it is also the inspiration for them all is an

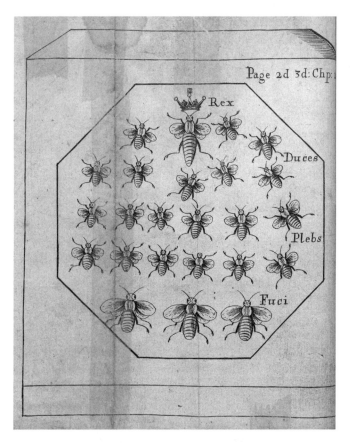

Figure 7

Moses Rusden, *A Further Discovery of Bees* (1679). Courtesy the Wellcome Collection.

open question, but it is in any case so rich in possibilities that it has thus far outstripped human inventiveness. We are, after all, only one species among millions, and for evidence of sheer baroque variety, natural history beggars cultural anthropology. And then there is all of inorganic nature to boot, from the orbits of planets to the symmetry of ice crystals. Nature displays so many kinds of order that it is a beckoning resource with which to instantiate any particular one imagined by humans. A resource can become a temptation when nature is imbued with superhuman authority, either as divine edict or sublime paragon—greater, grander, infinitely more powerful and enduring than anything human. In such cases, nature functions as justification as well as simple representation.

None of this makes the doubling of moral orders by natural orders inevitable. Steam engines, switchboards, and other human contrivances, not just cells and the solar system, have all at one time or another also served to model moral orders. We have seen how seventeenth-century natural philosophers such as Robert Boyle used clockwork to model nature itself. Nor does appeal to a natural order as model of a moral order necessarily fortify the latter with additional authority. For one thing, not all traditions draw a sharp ontological distinction between natural and moral realms. Anthropologist Philippe Descola points out that what he calls Western "naturalist" societies are historically and culturally unusual in insisting on such a categorical (and wildly asymmetric) distinction between the human and nonhuman aspects of the universe; other traditions freely interweave elements of the natural and moral orders in ways that naturalists (in Descola's sense) can only describe as "anthropomorphic" or "projective."[1]

Even to formulate the reproach of anthropomorphism implies a certain commitment to anthropocentrism. Only from a parochial human point of view does it make any sense to divide up all that exists into our species on one side and everything else, from microbes to pulsars, on the other. (Imagine such a division from the standpoint of some other species—raccoons, say—and the oddity of a universe split up between raccoons and not-raccoons becomes absurdly apparent.) And only once such a division is posited is it possible to identify anthropomorphic projections from the tiny province of the human onto the vast realm of the universe— and to declare them illicit or childish. Other cultures, even those in the Greco-Roman lineage, have divided up the world differently. When, for example, the ancient Greek philosopher Heraclitus coupled the due measure of the sun with that of justice, he may not have been speaking metaphorically: "The sun will not overstep his measures; otherwise the Erinyes, ministers of Justice, will find him out."[2] The measures of the sun and those of human justice belonged to the same realm, with no need of a metaphor to bridge them.

Nor do cultures that *do* draw a clear line between the natural and the human always accord greater majesty or dignity to nature. In early modern Europe, for example, human "civilization" was regularly and favorably opposed to "savage" nature. Human labor was thought to improve or perfect nature, and well-kept gardens were regarded as self-evidently preferable to wilderness, just as raw materials refined by human art—silver extracted from ore or linen woven from flax—were obviously more valuable. To cultivate land (or leave it uncultivated) could be grounds for granting or rescinding property rights, according to seventeenth-century philosopher John Locke: "*As much land* as a man tills, plants, improves, cultivates, and can use the product of, so much is his *property*."[3] Human nature itself was likened to a garden and the culture of the soul to that of the earth. Upbringing bore an analogy to watering and weeding, wrote the statesman and natural philosopher Francis Bacon: "A man's nature runs either to herbs or weeds; therefore let him seasonably water the one, and destroy the other."[4] Vigilance was required to keep from slipping back from a civilized into a natural state. One late-seventeenth-century English observer took the German word *Handschuh* ("glove," literally "hand-shoe") as evidence that the Germans had only recently given up walking like animals on all fours.[5]

Even cultures that do not categorically distinguish the natural from the human or, if they do, consider nature to be the worse side of the bargain, use aspects of the natural order to figure the moral order. Perhaps other sorts of intelligences, with different bodies and senses, or no bodies at all, would not need to figure anything. For Martians and angels, order might just be, requiring no representation. But for our species, with our sensorium, orders must be grasped and imagined, both literally and figuratively. Nature is abundantly available, opulently diverse, and orderly in every possible sense of the word. It is therefore not so surprising that nature should be quarried for the figures of order, any and all orders. The human impulse to make nature meaningful is rooted in a double insight about order: normativity demands order; and nature supplies exemplars of all conceivable orders.

But natural order alone cannot dictate which specific norms to follow, if only because there are so many orders in nature. Nature is every bit as fertile in variety as culture. Therefore, the hope that norms extracted from nature will converge more convincingly than those freely invented by art is illusory. In other words, the strategy of naturalization to combat relativism is doomed. To glorify certain human values as "natural," whether in the liberal cause of human rights or the conservative one of social Darwinism, does not lend them one iota more of certainty or inevitability. Opponents can always retort, "Which nature?" and counter with examples of another order, equally natural, to support the opposite position.

The very variety of natural orders nonetheless suggests why we cannot do without natural orders when we conceive of moral orders. Nature is a repository of all imaginable orders. This is why the word "nature" is so embarrassingly rich in definitions. There are specific natures (the nature of maple trees, the nature of salamanders, the nature of salt crystals); there are local natures (the tropics and the tundra, the lush valleys and the bald mountain peaks); there are universal natures (fire burns everywhere, zero degrees

Kelvin is absolute zero even in the remotest galaxy)—and these are no doubt only three among many possible natural orders. It is, so to speak, in the very nature of the word "nature" to mean many things. Therefore, whichever norms are drawn from one sense of nature are more than likely to be in competition with, if not contradiction to, other norms drawn from nature. It was just this proliferation of norms in nature that led critics like Mill to throw up their hands in exasperation: nature will never speak with one voice, so why listen?

The polyphony of nature is, however, precisely the point: it is difficult—perhaps impossible—to imagine an order that is not manifestly, flamboyantly on display in nature. Nature is that delightful paradox, a disorderly *Wunderkammer* of all possible orders. The Renaissance *Wunderkammer*, ancestor of modern museums, dramatized the fecundity and plenitude of nature in its floor-to-ceiling displays, juxtaposing flies in amber and stuffed crocodiles, two-headed cats and striped tulips, magnets and petrified wood in order to overwhelm the spectator with the glorious hodge-podgery of it all (fig. 8).[6] Although the *Wunderkammer* systematically favored the rare and the singular over the commonplace and the ordinary, even everyday nature overflows with variety; the surprises of ethnography (fancy thinking that!) pale beside those of natural history (fancy *being* that!). All human dreams of order, revolutionary or reactionary, local or global, are ultimately figured, made vivid and alluring, in nature's *Wunderkammer* of possible orders. Nature's teeming bounty, however, diverges from

[margin note, handwritten:] cabinet of curiosities

Figure 8
Wunderkammer. Ferrante Imperato, *Dell'historia naturale* (1599).

that of the *Wunderkammer* in one crucial respect: even at its most intricate and improbable, nature exhibits some kind of order. The *Wunderkammer* aimed to astonish by defying all expectations; nature is in contrast the source of all expectations. And without well-founded expectations, the world of causes and promises falls apart. Normativity, like nature, does not require a unique order, but it does require *some* order.

8 CONCLUSION: SAVING THE PHENOMENA

These humanly rational propensities have to do with the kind of organisms that we happen to be. We are outfitted with senses that convey the surfaces of things. Even when intellectual curiosity and technological ingenuity makes possible anatomy, geometry, the microscope, X-rays, and other ways of peering beneath surfaces, our way of probing the viscera of the world is to turn them into yet more surfaces. If by some miracle noumena, things-in-themselves, were revealed to us, we could only grasp them as phenomena, as appearances. Fortunately, the peculiarities of our sensory systems have not blocked philosophical and scientific inquiry into domains ordinarily inaccessible to the senses, from elementary particles to distant stars to brain waves. But even in these investigations there has been a strong tendency to convert information, much of it now digital, into appearances, especially images—from radio telescopes, bubble chambers, magnetic resonance scans, and innumerable other devices designed to penetrate where the senses cannot reach (fig. 9). When Plato attempted to wean his readers from their addiction to appearances in the *Republic*, the only way he could make his point was to invent a myth about more appearances: the shadows in the cave cast by the equally phenomenal, if occluded, objects outside in the daylight. For beings like us, it is appearances all the way down.[1]

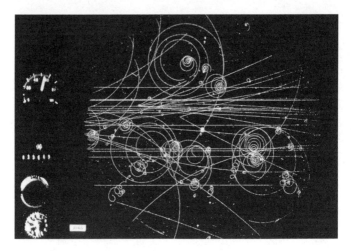

Figure 9
Trajectories of elementary particles in a bubble chamber, CERN, CERN-EX-11465-1 (1960–2017).

We are not content to receive appearances; we want to make them as well. To be fully real for us, a thing must appear, and this imperative holds for the very real moral orders devised by humans as well as for the artifacts they paint, build, mold, and forge. How to make a moral order appear? In principle, any surface, natural or artificial, will do, and there are examples of both sorts of models, palpable to the feeling as to sight: the ideal society as beehive or as clockwork. But in practice, it is natural surfaces that preponderate in our experience, by their sheer diversity, durability, and inescapability. The surfaces that nature presents so abundantly and incessantly to view are also ordered, in ways more obvious, more reliable, and more permanent than most artifacts. It is the natural appearances of day-in, day-out experience, not the natural depths revealed by electron microscopes and cyclotrons, which still shape some of our most sturdy intuitions about what an order can be. And there is no morality that does not conjure up an order, a bulwark against chaos. This is why natural phenomena (in the root sense of the word, "appearances") supply the material analogies closest to hand when human beings attempt to realize moral orders. This is the deep anthropological truth of myths of incarnation, the word made flesh.

We are now finally in a position to return to Kant's deprecating remarks about the limits of our abilities to imagine rational beings who are truly inhuman. Kant's ironies about our provincialism in this regard echo the scoffing of Xenophanes of Colophon (570?–475? BCE), who ridiculed the provincialism of religion: "The Ethiopians say that their gods are snub-nosed and black, the Thracians that theirs have light blue eyes and red hair. / But if cattle and horses or lions had hands, or were able to draw with their hands and do the works that men can do, horses would draw the form of the gods like horses, and cattle like cattle, and they would make their bodies such as they each had themselves."[2] The skeptics will cite such passages to press their point: even if the appeal to natural order to ground any and all norms is an ineradicable aspect of the kind of beings we are, a genuine feature of our species anthropology in the philosophical sense, isn't this a matter for regret? Is our appeal to natural order to ground norms made one whit less irrational (dangerous, even) by being universal? Much of the urgency of this question comes from the understandable fear that grounding norms in nature can lead to an unthinking conservatism: if the norms come from nature, and nature is unchanging, then so are the norms. I would offer three concluding responses to those skeptics who still insist that the naturalistic fallacy deserves to be called such.

1. Naturalization is in fact a weaker strategy than its critics fear: there are natural orders aplenty to support (or subvert) any and all norms.

First, the shift from using this or that natural order to justify a particular set of norms to using some kind of natural order to justify any set of norms draws a great deal of the political sting of what philosophers have objected to as specific instances of the naturalistic fallacy. Once one realizes that there is no one unique order of nature in which to ground norms, the force of any one such appeal weakens dramatically. For any example from natural history I can come up with to support my favorite norms, you can come up with a plenitude of other, equally natural analogies to support very different norms: the matriarchy of bees versus the patriarchy of baboons. Nature is no longer a knockdown weapon in a political argument, because all parties to a controversy can wield it.

2. The appeal to nature is fundamentally about the link between natural order and normativity per se, not the link between any particular natural order and any particular set or norms.

This brings me to my second response: the move to appeal to natural order in order to support visions of the moral order is fundamentally about the link between order and normativity per se, not about the link between any specific order and any specific set of norms. Once again, it is the presumption of uniqueness that makes the appeal to nature fallacious, not the presumption of order. I have argued that the connection between order and normativity is a necessary one. The link between *natural* order and normativity is a contingent one, but it is extraordinarily strong and tenacious, even in the largely built world of late modern societies. It would be worth further inquiry as to why the natural order sometimes—not always—provides the resources to justify as well as to represent moral order. In addition to the religious and metaphysical traditions already mentioned, there is the fact that the sheer scale and durability of nature have surpassed that of even the most impressive human artifacts for most of human history. Natural orders are, in effect, more orderly than human orders, which may offer a clue as to why natural orders are invoked to buttress human orders and not vice versa. In an age of genetic engineering and anthropogenic climate change this imbalance of power may be shifting in the opposite direction.

3. Human reason in human bodies is the only kind of reason we have.

Third and finally, it is pointless to yearn for that which is in principle unattainable. Human reason in human bodies is the only kind of reason we have. As I have argued, the specifics of the human sensorium have profoundly shaped our cognition, particularly our drive to represent everything from society to digital data in appearances, whether in the form of beehives or images of remote galaxies. The yearnings of philosophers for another kind of reason, allegedly more perfect, have always been enmeshed in theology, either overtly or covertly. It is no accident that qualms about anthropomorphism and idolatry, of which Xenophanes's acerbic remarks are an early example, first occur in the context of religion. They ridicule (or censure) the admixture of the human with the divine. The earliest bans on anthropomorphism in science come much later, in the seventeenth-century works of Francis Bacon, René Descartes, and others. But they too recur repeatedly to metaphors of idolatry and dreams of angelic and divine intelligences more perfect than our own, with different bodies and senses or no bodies and senses at all. Epistemology still indulges in thought experiments in angelology, whether in Kant's mysterious nonhuman rational beings or Martians or the inhabitants of other possible worlds. Theology continues to haunt epistemology, feeding desires that can never be realized for a form of reason that escapes the limitations of our species. Kant famously warned against the ambitions of reason to transcend its own limitations. Perhaps we might follow Kant in spirit if not in letter if we explored the capacities of specifically *human* reason.

Notes

CHAPTER 1

1. Immanuel Kant, *Anthropology from a Pragmatic Point of View* (1798), trans. and ed. Robert B. Louden (Cambridge: Cambridge University Press, 2006), I.30, 65.

2. Kant once wrote (in the context of how to weigh the strength of belief by probabilities) that he would be willing to bet all he had on the existence of life on other planets: "I should be willing to stake my all on the contention—were it possible by means of any experience to settle the question—that at least one of the planets we see is inhabited. Hence I say that it is not merely opinion, but a strong belief, on the correctness of which I should be prepared to run great risks, that other worlds are inhabited." Immanuel Kant, *Critique of Pure Reason*, trans. Norman Kemp Smith (New York: St. Martin's Press, 1965), A825/B823, 648.

3. For examples, see William Cronon, ed., *Uncommon Ground: Rethinking the Human Place in Nature* (New York: Norton, 1996); Mikulás Teich, Roy Porter, and Bo Gustafsson, eds., *Nature and Society in Historical Context* (Cambridge: Cambridge University Press, 1997); Lorraine Daston and Fernando Vidal, eds., *The Moral Authority of Nature* (Chicago: University of Chicago Press, 2004); and the still fundamental Clarence J. Glacken, *Traces on the Rhodian Shore: Nature and Culture in Western Thought from Ancient Times to the End of the Eighteenth Century* (Berkeley: University of California Press, 1967).

4. The British philosopher G. E. Moore first coined this term in the context of ethics: G. E. Moore, *Principia Ethica* (1903; Cambridge: Cambridge University Press, 1976), 37–58. Since

then, the term's range of references has expanded to include any appeal to nature as a standard for human values: see Lorraine Daston, "The Naturalistic Fallacy Is Modern," *Isis* 105(2014): 579–587.

5. Friedrich Engels to Pjotr Lawrowitsch Lawrow, November12–17, 1875, in Karl Marx and Friedrich Engels, *Werke* (Berlin: Dietz Verlag, 1966), vol. 34, 170.

6. John Stuart Mill, "Nature," in *Three Essays of Religion* (1874) in Mill, *Essays on Ethics, Religion and Society*, ed. J. M. Robson (London: Routledge, 1996), 373–402, on 386.

CHAPTER 2

1. See, for example, Arthur O. Lovejoy, "'Nature' as Aesthetic Norm," in Lovejoy, *Essays in the History of Ideas* (Baltimore: Johns Hopkins University Press, 1948), 69–77; Raymond Williams, *Keywords: A Vocabulary of Culture and Society*, rev. ed. (New York: Oxford University Press, 1985), 219–224.

2. Harald Patzer, "Physis. Grundlegung zu einer Geschichte des Wortes," *Sitzungsberichte der wissenschaftlichen Gesellschaft an der Johann Wolfgang Goethe-Universität Frankfurt am Main* 30 (1993): 217–280. See also R. G. Collingwood, *The Idea of Nature* (Oxford: Oxford University Press, 1960), 43–44; William Arthur Heidel, "Peri Physeos: A Study of the Conception of Nature among the Pre-Socratics," *Proceedings of the American Academy of Arts and Sciences* 45 (1910): 79–133, esp. 97–99.

3. See, for example, the entries "Nature" in the *Oxford English Dictionary*, "Natur" in *Grimms Wörterbuch*, and "Nature" in *Le Robert: Dictionnaire historique de la langue française*.

4. Wendy Doniger O'Flaherty, *The Origins of Evil in Hindu Mythology* (Berkeley: University of California Press, 1976), 94–95.

5. Scott Atran and Doug Medin, *The Native Mind and the Cultural Construction of Nature* (Cambridge, MA: MIT Press, 2008), 20–21.

6. Aristotle, *Parts of Animals*, trans. A. L. Peck, Loeb ed. (Cambridge, MA: Harvard University Press, 1998), I.i, 641b25–32, 73.

7. Aristotle, *The Physics*, trans. Philip H. Wicksteed and Francis M. Cornford, Loeb ed. (Cambridge, MA: Harvard University Press, 1980), II.i, 193b9–10, 115.

8. Aristotle, *Politics*, trans. H. Rackham, Loeb ed. (Cambridge, MA: Harvard University Press, 1990), I.iii.23, 1258b6–8, 51.

9. Aristotle, *Generation of Animals*, trans. A. L. Peck, Loeb ed. (Cambridge, MA: Harvard University Press, 1963), IV.iii, 767b9–10, 401.

10. Arnold I. Davidson, "The Horror of Monsters," in *The Boundaries of Humanity: Humans, Animals, Machines*, ed. James J. Sheehan and Morton Sosna (Berkeley: University of California Press, 1991), 16–67. Aristotle believed that interspecies crosses could result in offspring only if sizes and gestation periods were similar and gives several examples: Aristotle, *Generation of Animals*, II.vii, 746a27–746b6, 243–245.

11. Aristotle, *The Physics*, II.viii, 199b27–29, 179.

12. Immanuel Kant, *Critique of Pure Reason* (1781, 1787), trans. Norman Kemp Smith (New York: St. Martin's Press, 1965), A100–101, 132. Kant bases his argument for the necessity of the transcendental faculty of the imagination on this thought experiment of a world without specific natures.

CHAPTER 3

1. Herodotus, *The History*, trans. David Grene (Chicago: University of Chicago Press, 1987), II.35–36, 145–146.

2. *Airs Waters Places*, in *Hippocrates*, trans. W. H. S. Jones, Loeb edition, 2 vols. (1923; Cambridge, MA: Harvard University Press, 1995), vol. 1, 70–137, on 105–107.

3. Herodotus, *The History*, III.106–109, 257–258.

4. *Airs Waters Places*, 115.

5. *Airs Waters Places*, 111.

6. Carolus Linnaeus, *Oeconomia naturae* (Uppsala: Isaac Biberg, 1749). On the reception of Linnaeus's "economy of nature," see Donald Wooster, *Nature's Economy: A History of Ecological Ideas* (1977; Cambridge: Cambridge University Press, 1985), 31–49.

7. James Lovelock, *The Revenge of Gaia: Why the Earth Is Fighting Back—and How We Can Still Save Humanity* (London: Penguin, 2006), 16.

8. Lovelock, *Revenge of Gaia*, 26–37.

CHAPTER 4

1. Seneca, *Naturales quaestiones*, trans. Thomas H. Corcoran (Cambridge, MA: Harvard University Press, 1971), 2 vols., vol. 2, 276–281, VII.25; Brad Inwood, *Reading Seneca: Stoic Philosophy in Rome* (Oxford: Oxford University Press, 2005), 232. On laws of nature more generally in ancient philosophy, see Daryn Lehoux, "Laws of Nature and Natural Laws," *Studies in History and Philosophy of Science Part A* 37 (2006): 527–549.

2. Ian Maclean, "Expressing Nature's Regularities and their Determinations in the Late Renaissance," in *Natural Laws and Laws of Nature in Early Modern Europe: Jurisprudence, Theology, Moral and Natural Philosophy*, ed. Lorraine Daston and Michael Stolleis (Farnham: Ashgate, 2008), 29–44; Jane Ruby, "The Origins of Scientific Law," *Journal of the History of Ideas* 47 (1986): 341–359.

3. In addition to the essays in Daston and Stolleis, eds., *Natural Laws and Laws of Nature in Early Modern Europe*, see John Milton, "The Origin and Development of the Concept of the 'Laws of Nature,'" *Archives Européennes de Sociologie* 22 (1981): 173–195; John Henry, "Metaphysics and the Origins of Modern Science: Descartes and the Importance

of Laws of Nature," *Early Science and Medicine* 9 (2004): 73–114; Sophie Roux, "Les lois de la nature à l'âge classique: La question terminologique," *Revue de Synthèse* 4 (2001): 531–576; and Friedrich Steinle, "The Amalgamation of a Concept: Laws of Nature in the New Sciences," in *Laws of Nature: Essays on the Philosophical, Scientific, and Historical Dimensions*, ed. Friedel Weinert (Berlin: Walter De Gruyter, 1995), 316–368.

4. See the articles by Catherine Wilson, Ian Maclean, Gerd Graßhof, Sophie Roux, Jean Armogathe, and Friedrich Steinle, in Daston and Stolleis, eds., *Natural Laws and Laws of Nature in Early Modern Europe*.

5. Walter Cahn, *Masterpieces: Chapters on the History of an Idea* (Princeton: Princeton University Press, 1979), 90–91.

6. Robert Boyle, *A Free Inquiry into the Vulgarly Received Notion of Nature* (ca. 1666), in *The Works of the Honourable Robert Boyle* (1772), ed. Thomas Birch, 6 vols. (Hildesheim: Georg Olms, 1966), vol. 5, 158–254, on 164.

7. Boyle, *Free Inquiry*, 163.

8. Boyle, *Free Inquiry*, 188. The issue of idolatry exercised Leibniz as well, who wrote a reply to the Latin version of Boyle's *Free Inquiry*: Catherine Wilson, "*De ipsa natura*: Leibniz on Substance, Force, and Activity," *Studia Leibnitiana* 19(1987): 148–172, and more generally on the idolatry debate in early modern natural philosophy, Martin Mulsow, "Idolatry and Science: Against Nature Worship from Boyle to Rüdiger, 1680–1720," *Journal of the History of Ideas* 67(2006): 697–711.

9. On the shifting relationship between the terminology of "laws" and "rules" in seventeenth-century natural philosophy, see Friedrich Steinle, "From Principles to Regularities: Tracing 'Laws of Nature' in Early Modern France and England," in Daston and Stolleis, eds., *Natural Laws and Laws of Nature in Early Modern Europe*, 215–232.

10. H.G. Alexander, ed., *The Leibniz–Clarke Correspondence* (New York: Manchester University Press, 1998).

CHAPTER 5

1. See, for example, the protestors pictured in "Should We Allow Research on Human-Animal Embryos?" *Guardian*, January 12, 2009.

2. See, for example, "Nature's Revenge," *New York Times*, August 30, 2005, apropos of Hurricane Katrina, or "Mother Nature's Revenge against Human Development," *Independent*, October 24, 2007, apropos of California wildfires.

3. Thomas Aquinas, *Summa contra gentiles*, trans. Vernon J. Bourke, 3 vols. (Notre Dame: University of Notre Dame Press, 1975), 3.101.2, vol. 3, pt. 2, 82.

4. Lorraine Daston and Katharine Park, *Wonders and the Order of Nature, 1150–1750* (New York: Zone Books, 1998), 303–328.

5. Philip Fisher, *The Vehement Passions* (Princeton: Princeton University Press, 2002), 43.

6. Fisher, *Vehement Passions*, 44.

7. Immanuel Kant, *Grundlegung zur Metaphysik der Sitten* (1785), ed. Theodor Valentiner (Stuttgart: Reclam, 2000), 58.

CHAPTER 6

1. Thomas Henry Huxley, "Evolution and Ethics," *Evolution and Ethics and Other Essays* (London: Macmillan, 1894), 46–116, on 83.

2. Christine M. Korsgaard, *The Sources of Normativity*, ed. Onora O'Neill (Cambridge: Cambridge University Press, 1996).

3. G. S. Kirk and J. E. Raven, *The Presocratic Philosophers. A Critical History with a Selection of Texts* (Cambridge: Cambridge University Press, 1969), 225–227.

4. On the role of nature in the Abrahamic religions, as well as in ancient Greece and Rome, see Rémi Brague, *La Sagesse du monde* (Paris: Fayard, 1999).

5. Aristotle, *On the Heavens*, trans. W. K. C. Guthrie, (Cambridge, MA: Harvard University Press, 1971), I.iii, 270b1–12, 24–25. See also fragment 224 from Heraclitus: "The ancients assigned to the gods the heaven and the upper region as being the only immortal place ... " (translation from Kirk and Raven, *Presocratic Philosophers*, 200).

6. Ian Hacking, *Representing and Intervening: Introductory Topics in the Philosophy of Natural Science* (Cambridge: Cambridge University Press, 1983), 132.

CHAPTER 7

1. Philippe Descola, *Par-delà de la nature* (Paris: Gallimard, 2005), 101–107.

2. Heraclitus, fragment 229, as translated in G. S. Kirk and J. E. Raven, *The Presocratic Philosophers: A Critical History with a Selection of Texts* (Cambridge: Cambridge University Press, 1969), 203. On the significance of measure in ancient Greek thought, see Laura M. Slatkin, "Measuring Authority, Authoritative Measures: Hesiod's *Works and Days*," in *The Moral Authority of Nature*, ed. Lorraine Daston and Fernando Vidal (Chicago: University of Chicago Press, 2004), 25–49.

3. John Locke, *Second Treatise of Government* (1690), ed. C. B. Macpherson (Indianapolis: Hackett, 1980), 5.25, 21.

4. Francis Bacon, "Of Nature in Men," *Essays Civil and Moral*, in *Works*, in *Lord Bacon's Works*, ed. Basil Montagu, 16 vols. (London: William Pickering, 1825–34), vol. 1, 135.

5. Keith Thomas, *Man and the Natural World: A History of the Modern Sensibility* (New York: Pantheon, 1983), 132–133.

6. Lorraine Daston and Katharine Park, *Wonders and the Order of Nature, 1150–1750* (New York: Zone Books, 1998), 255–276.

CHAPTER 8

1. There have been poets and philosophers of surfaces as well as depths: see Wendy Doniger, *The Woman Who Pretended to Be Who She Was: Myths of Self-Imitation* (Oxford: Oxford University Press, 2005), 213–214.

2. G. S. Kirk and J. E. Raven, *The Presocratic Philosophers: A Critical History with a Selection of Texts* (Cambridge: Cambridge University Press, 1969), 168–169.